# Square roots
# of Numbers

# Square roots of Numbers

## Prime elimination game

# Epsilon The Heir

| Library of Congress Control Number: | | 2022901212 |
|---|---|---|
| ISBN: | Hardcover | 978-1-6698-0808-4 |
| | Softcover | 978-1-6698-0807-7 |
| | eBook | 978-1-6698-0806-0 |

Print information available on the last page.

Rev. date: 01/24/2022

**To order additional copies of this book, contact:**
Xlibris
844-714-8691
www.Xlibris.com
Orders@Xlibris.com
834978

$$N\sqrt{2}$$

# WELCOME TO PRIME SQUARE ROOT OF NUMBERS

# Contents

# MEANINGS AND DEFINITIONS

$N\sqrt{2}$   N = any number        $\sqrt{2}$ = square root

Potential Prime Numbers = Numbers that can not be divided by 2, 3 and 5.
Prime = numbers that can only be divided by 1 and the number itself.
Prime number chart = Prime numbers 7 and up.
Lowest Common Denominator(s) = Prime numbers 2, 3, and 5
Highest Common Denominator(s) = Prime numbers 7 and up.
Potential Prime number sequence(s) = numbers that cannot be divided by 2, 3, and 5.
Pre-set value = fixed for any input.
1, 2, 3, 4, 5... = numbers or number scale.
Number(s) = 1, 2, 3, 4, 5...any number (s)

# INTRODUCTION

FOR HUNDREDS OF years, the greatest minds have had the pleasure to toy with Prime.

Welcome to Square root of Numbers Prime. Numbers are so interesting a person could write almost forever, just about numbers itself. Prime alone would take half the time.

Numbers are not a creation of humans, nor is it a Human invention. Numbers are devised or deciphered by Humas. The only one way numbers could come into Existence was right from the beginning of creation. Right from the start Go, one was never without the other. The Beginning was never without Numbers.

In the next few pages will be more insights of Numbers, plus Prime solved and made easy to use. Hopefully it is straight forward and easy to understand, if not, please offer feedback for future simplification and clarity. Please enjoy, with thanks

FREDI

Prime is simply a wise persons game, by removing number 1 from use 2, 3, and 5 are exceptional Prime numbers and are the only lowest common denominator that can be simply removed from the number scale totally. When all common denominators of 2, 3, and 5 are removed, we are left with very few numbers.

When 2, 3 and 5 and all the common denominators that go with these three numbers are removed, the remaining numbers follow a sequence of 30 numbers.

2, 3, 5, and 10 together the repeat every 30 numbers. Diagram 2, take a look at these 4 numbers that repeat every 30 numbers, the number they can not make is 1, 7, 11, 13, 17, 19, 23, and 29 the sequences follow this order by 30 numbers every 30 numbers.

# DIAGRAM I:

| | | | 0 | 1 | 2 | 3 | 4 | 5 | 6 | 7 | 8 | 9 |
|---|---|---|---|---|---|---|---|---|---|---|---|---|

Numbers are arranged in sequences of 10's, then they simply repeat by adding the numbers of 10's to it. In simple, 0, 1, 2, 3, 4, 5, 6, 7, 8, and 9 these last numbers keep repeating over and over again, numbers are grouped in 10's. Each number is not group in 10's meaning that some of the numbers like 1, 3, 4, 6, 7, 8 and 9 can not make its own numbers as it goes along. Example, 3 can not make 13 and 23 or 4 can not make 14 and 34. Other numbers are left to make numbers which can not be made by themselves. Number 2 is a special number that can make all even number that can not make themselves and making it the only even Prime numbers. Once we remove all even numbers, we are left with odd numbers only.

All Prime numbers end with these numbers after the number these are 1, 3, 7, and 9 all the time and end with no other numbers. There is a reason for this 1, 3, 7, and 9 can not make their own numbers in groups or sequences of 10's as it goes along the number scale.

# DIAGRAM 2:

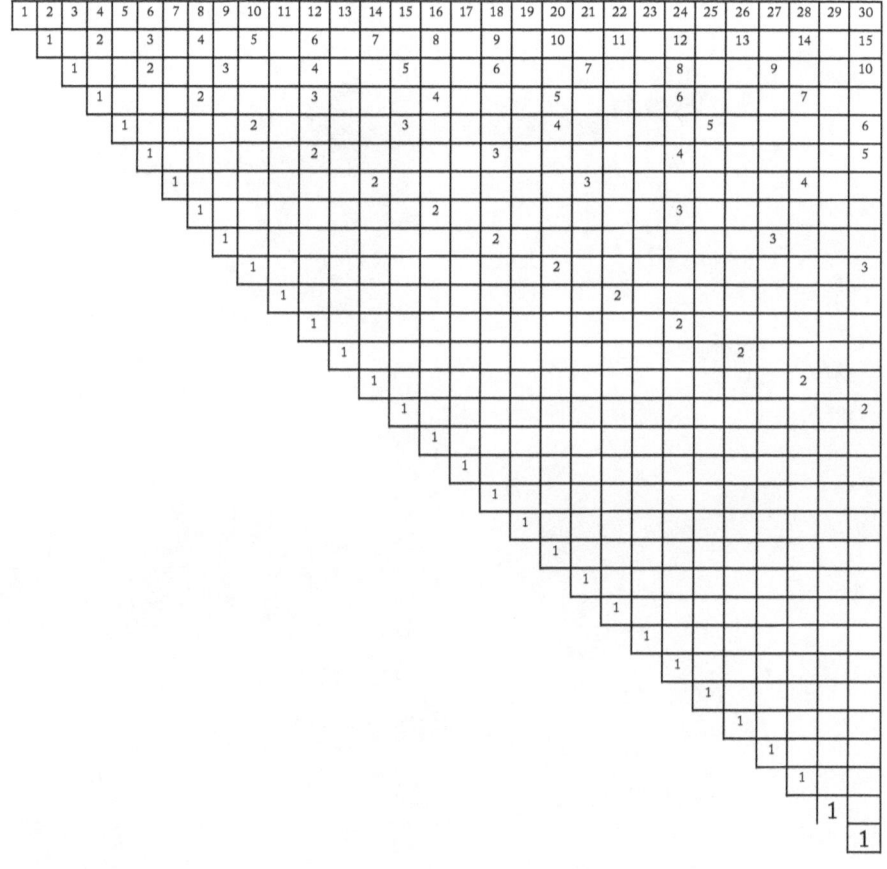

Division
Graft

(sequence of 30 numbers)

Number 1 is the Hypotenuse to all numbers

Each number has its own value of wavelength. Number one is the master key to all numbers. Addition, subtraction, division and multiplication are different Mandalas (Grafts). Numbers have a pre-set value. Whether we set the value, no value has been set or value is unknown. In each case, numbers are always the same. Numbers had to exist right from the beginning of All Things, with a "Pre-set Values".

In division, numbers run parallel in ratio with number one on the hypotenuse, the number in questions ratio is always over one. All numbers square with number one.

When numbers meet at 90 degree on a right angle from number one, is never a prime number. Prime numbers are numbers that no two natural numbers meet on 90 degree from number one. Number one is the only number that meets all numbers on 90 degree. (See Diagram 2)

Numbers are amplitude and wavelength, each number has the same amplitude. The wavelength of each number is different when multiplying or dividing numbers, the answer will always be the wavelength.

Once number one is set, so are the rest of numbers. No matter what amplitude or wavelength number one is set at, the rest of numbers contain the same amplitude, plus the wavelength distance from number one.

The number itself tells the wavelength distance from number one.

All numbers have same amplitude as number one. Number one's wavelength times the number in question is the wavelength of a number. Numbers are different wavelength, the number itself is the wavelength of a number measured from number one in multiple of number one.

After everything said and done, numbers are educational and very fulfilling. Within the number system Prime has an extra special quality, as the Prime number becomes more larger, the more rare and unique it is. The larger the Prime number is, the more uniquely it is a "one of a kind". The smaller the Prime number is, the more it has duplicates.

Number 1 creates all number first that cannot be made by any other two number and these numbers are Prime numbers.

# POTENTIAL PRIME NUMBERS SEQUENCE CHART DIAGRAM 3

Removed

| 2 |
|---|
| 3 |
| 5 |

Lowest Common
Denominators
(Prime only)

| 7 | 37 | 67 | 97 | 127 | 157 | 187 | | | | | |
|----|----|----|-----|-----|-----|-----|---|---|---|---|---|
| 11 | 41 | 71 | 101 | 131 | 161 | 191 | | | | | |
| 13 | 43 | 73 | 103 | 133 | 163 | 193 | | | | | |
| 17 | 47 | 77 | 107 | 137 | 167 | 197 | | | | | |
| 19 | 49 | 79 | 109 | 139 | 169 | 199 | | | | | |
| 23 | 53 | 83 | 113 | 143 | 173 | 203 | | | | | |
| 29 | 59 | 89 | 119 | 149 | 179 | 209 | | | | | |
| 31 | 61 | 91 | 121 | 151 | 121 | 211 | | | | | |

ᴸ Numbers which are not in Common with lowest common Denominators (2, 3, and 5) (in groups of 30 numbers) ᴶ

C IRCULED NUMBERS ARE which have been eliminated by the Highest Common Denominators (chart diagram 4).

Prime shows what is not possible without the use of number 1. Only a higher common denominator (Prime). Diagram 4 can eliminate a Potential Prime Number. Any number that follows the sequences of Potential Prime numbers (Diagram 3) that has not been eliminated is a Prime Number.

# ELIMINATION OF POTENTIAL PRIME

## PRIME NUMBERS CHART

## DIAGRAM 4:

Removed

| 2 |
| 3 |
| 5 |

Lowest
Common
Denominators
(Prime only)

|     | 7   | 11  | 13  | 17  | 19  | 23  | 29  | 31  |   |   |   |
|-----|-----|-----|-----|-----|-----|-----|-----|-----|---|---|---|
| 7   | 49  | 77  | 91  | 119 | 133 | 161 | 203 | 217 |   |   |   |
| 11  | 77  | 121 | 143 | 187 | 209 |     |     |     |   |   |   |
| 13  | 91  | 143 | 169 | 221 |     |     |     |     |   |   |   |
| 17  | 119 | 137 | 221 | 289 |     |     |     |     |   |   |   |
| 19  | 133 | 209 | 247 |     |     |     |     |     |   |   |   |
| 23  | 161 | 253 |     |     |     |     |     |     |   |   |   |
| 29  | 203 |     |     |     |     |     |     |     |   |   |   |
| 31  | 217 |     |     |     |     |     |     |     |   |   |   |
|     |     |     |     |     |     |     |     |     |   |   |   |
|     |     |     |     |     |     |     |     |     |   |   |   |
|     |     |     |     |     |     |     |     |     |   |   |   |
|     |     |     |     |     |     |     |     |     |   |   |   |

# HIGHEST COMMON DENOMINATORS (PRIME ONLY)

## REPLACING NUMBER 1

S IMPLY TURNING PRIME into an elimination game. Prime can never be dissolved by the use of Highest Common Denominators (Prime). Prime are first time numbers that can replace number 1. The larger the number pool of Prime numbers the slimmer Prime becomes.

Prime numbers are the Higher Common Denominator replacing number 1 in its place. Each higher common denominator (Prime) is a New number 1 in the place of number 1 that has no commoness with any Prime numbers other than the ability to squaring with one another. Prime numbers only have one use to square with one another once. All numbers square with each other only once.

Prime will never stop. As soon as prime slows down or looks to be ending, then the more Prime are produced. There is a reason for this, if Prime stops then it cant produce numbers anymore. This will produce more Prime again to

make numbers. Prime numbers are the only numbers that produce numbers in replacement of number 1.

Only Prime number can eliminate Potential Prime numbers or only higher common denominator can eliminate a Potential Prime number.

# SUMMARIZATION

PRIME WITH THE use of Potential Prime number chart (Diagram 3) and Prime number chart (Diagram 4). To understand why Potential Prime number chart works, we need to start with the first three lowest common denominators 2, 3 and 5. The even number 2 is easy, it is simply removes all even, for it is the lowest common denominator to all even numbers on the number scale. Next, we have a number 3 and 5, to work with all odd numbers on the number scale. Number 5 is also easy, it simply removes all odd number 5 from the number scale. We only have number 3 left, number 3 removes everything number 9 can do. Now when we look at the number system, they are grouped in order of 10's. They only two numbers that could not be done with three (2, 3 and 5) common denominators was 1 and 7, we know why number 1 could not be done, we have removed number 1 from use. So number 7 was the only one that could not be made, keep this in mind for now. Numbering system is our next problem to solve, it is grouped in 10's. In this fashion 0, 1, 2, 3, 4, 5, 6, 7, 8, and 9 then it repeats over and over again the last numbers.

Each number works different than our number scale system. Example, number 3 repeats every third number 5 repeats every fifth number, each number repeats it own number only. Now we need to connect 2, 3, 5, and 10 together in a rhythm of 2, 3, 5 and 10, they keep repeating every thirty numbers. We look and see which numbers they cannot make. They are 1, 7, 11, 13, 17, 19, 23, 29. These last eight numbers keep repeating every 30 numbers. Simply change the 10's in front of each number as you go up the number scale by adding 30 to each number. We have now changed the number system to groups of 30's or the rhythm of 30's of all the numbers which could be a Potential Prime Number.

Only numbers left in the number system that can make numbers in

replacement of number 1 are Prime numbers only. Since we know now that Prime numbers become a higher common denominators as Prime numbers can only make extra numbers by squaring with one another.

By squaring Prime numbers, you can be sure that the only numbers they can produce are Potential Prime numbers and no other numbers. Now that we know this, we can draft a Prime Number Chart (Diagram 4). When we use this two charts (Diagram 3) and (Diagram 4), very little math or computation is required. Plus in addition, we can see in advance what number may or may not be Prime next.

Remember when asked to keep the number 7 in mind, the reason for this is, number 7 is the first highest common denominator that can make an extra number by squaring with itself. The first number it make is 49. Number 7 becomes the first permanent higher common denominator on (Diagram 4). All higher common denominators (Prime number) are permanent making new numbers by squaring with each New Prime numbers only. Next numbers on the chart are all Prime number only (Diagram 4) comes, 11, 13, 17, 19..and so on. Every time there is a new Prime numbers then placed on the chart to be squared and make a New Number.

Two charts are required to play;

1.  Potential Prime number Sequence Chart (Diagram 3)
2.  Prime Numbers or Highest Common Denominator Chart (Diagram 4)

With these two charts anyone and everyone can be a Prime Pro. Simply follow the steps. Fill in the empty spaces on Diagram 3 and 4 which have been kept empty for you to try. Make your own as large and long as you wish.

Start with Chart Diagram 3, simply add 30 to each number next to be placed on the chart horizontally to fill the chart. Then go to chart Diagram 4 and see which number is same on Diagram 3, circle that number because it is not a Prime. All numbers which have not been circled or eliminated are Prime numbers, once you know which are Prime numbers, take these numbers, from Diagram 3 and add or place them into Diagram 4 for elimination of Diagram 3. Keep doing this over and over again, you are a Prime Pro Now.

PRIME ELIMINATION
GAME
ANYONE CAN PLAY
ENJOY

# NOTES

# PLAY GRAFT

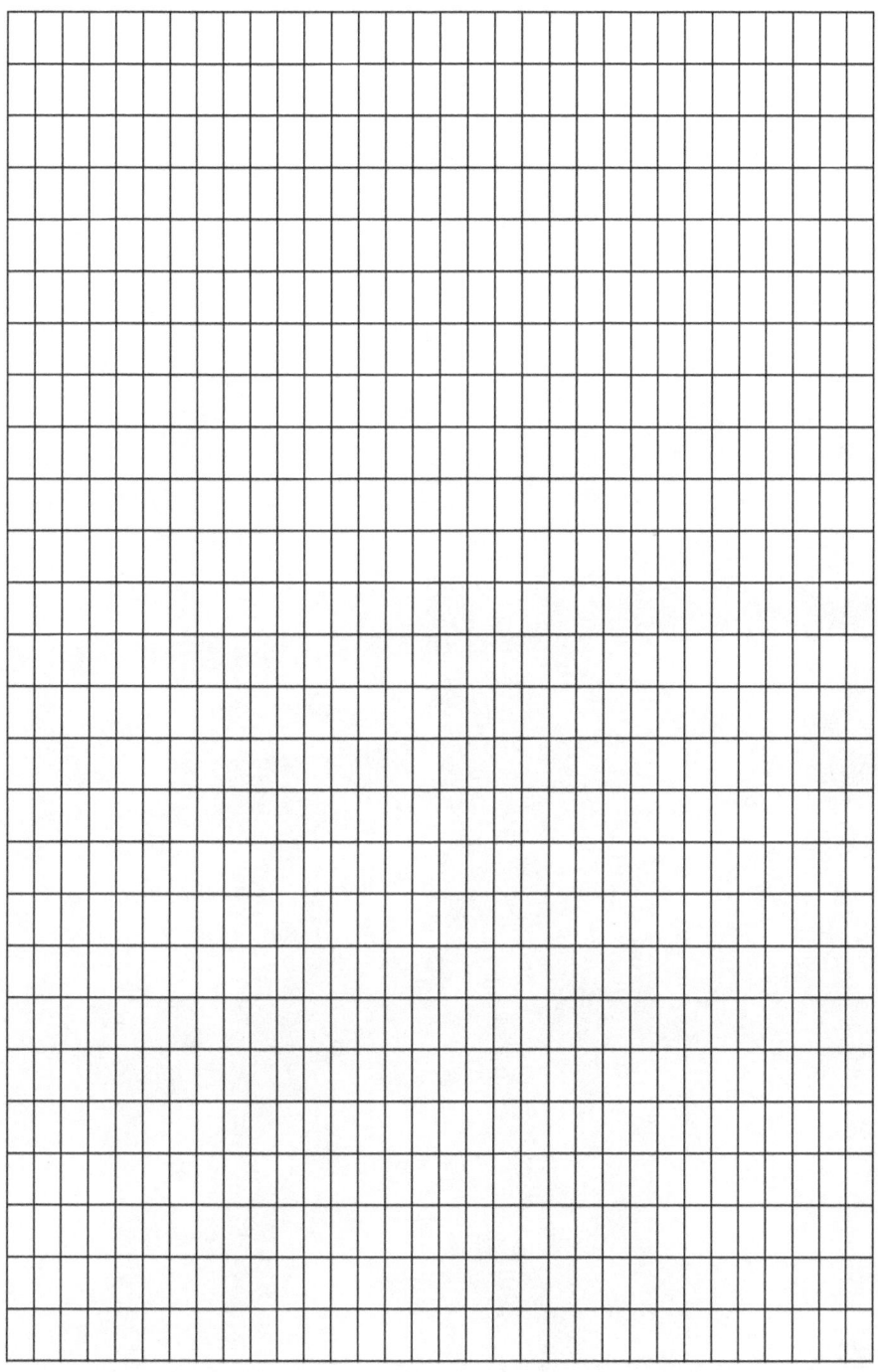

# PLAY GRAFT

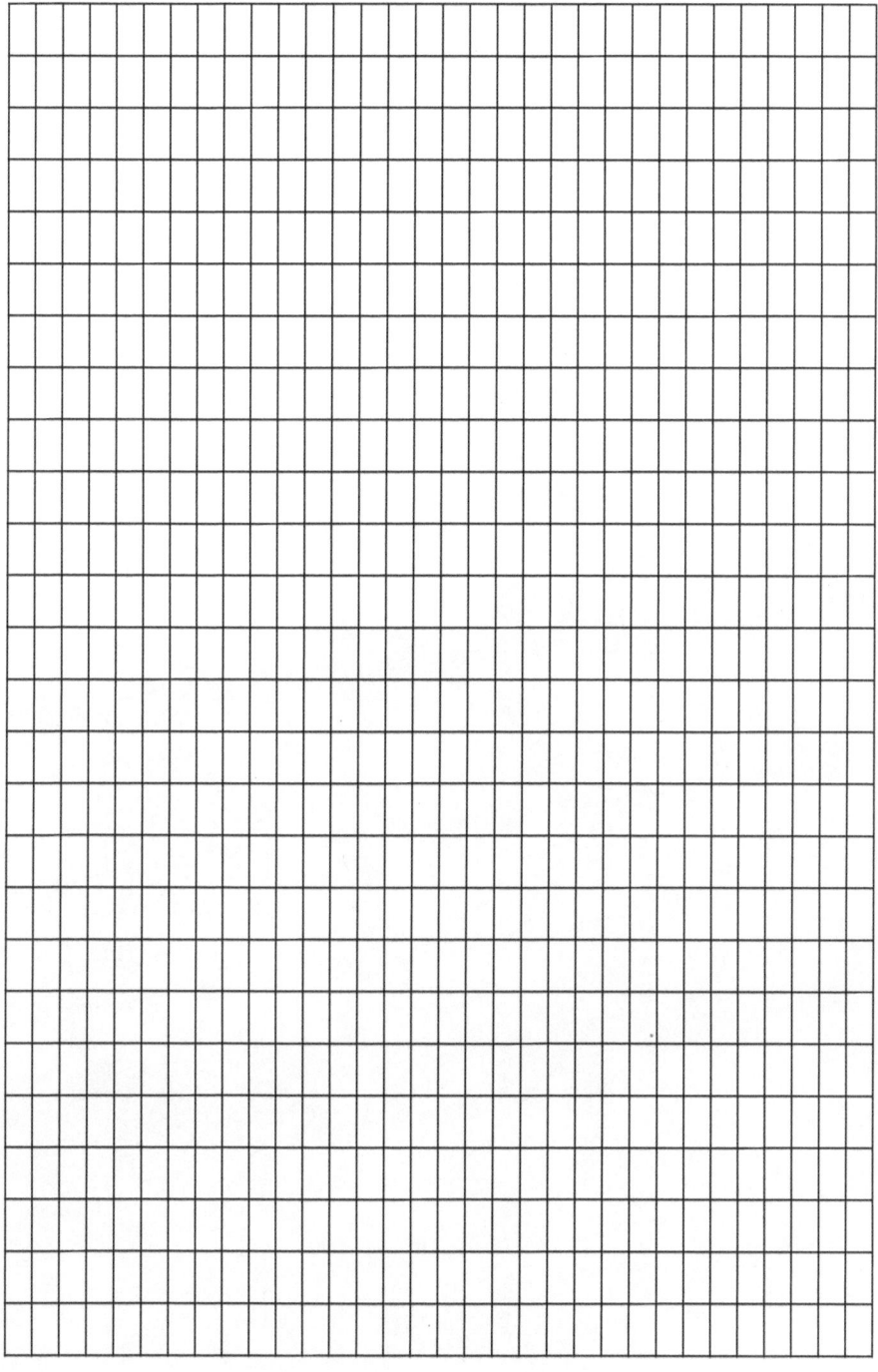

# PLAY GRAFT

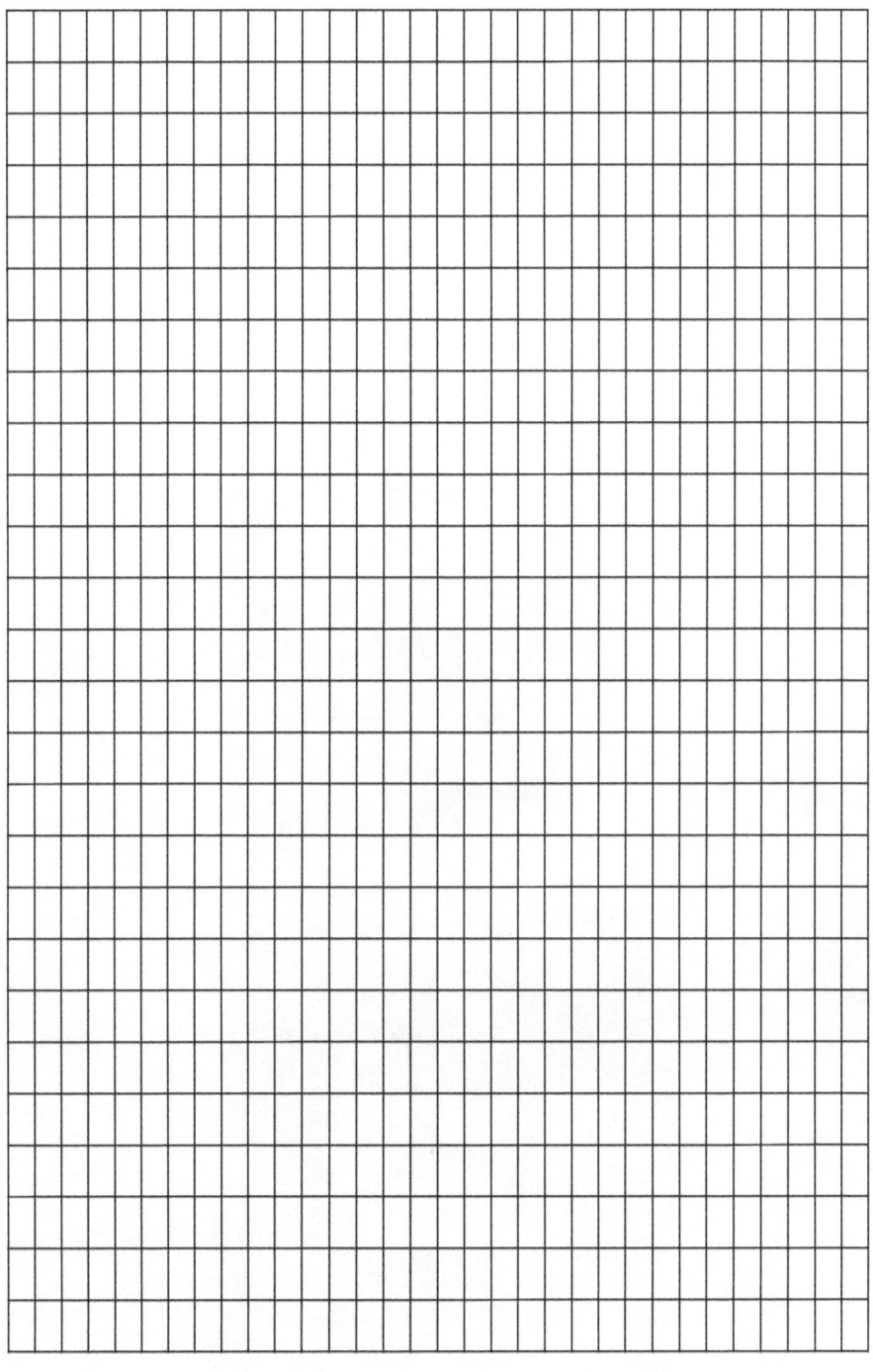

www.ingramcontent.com/pod-product-compliance
Lightning Source LLC
Chambersburg PA
CBHW031503210526
45463CB00003B/1053